HORSEFIELD TORTOISES

Horsefield tortoise care, health, diet, breeding, cages, pro's and cons and lots more included

Ben George Carre

Table of Contents

Introduction

Originally from the arid regions of Central Asia, Horsefield Tortoises, also known as Russian Tortoises, have distinct characteristics that make them unique and endearing companions. Welcome to the fascinating world of Horsefield Tortoises, where these captivating reptiles have captured the hearts of reptile enthusiasts and pet owners alike.

This in-depth guide explores the nuances of caring for a Horsefield Tortoise and offers crucial advice to guarantee your pet's health and happiness. From setting up the ideal environment and comprehending behavior to dealing with common health problems, we set out to discover the mysteries of conscientious Horsefield Tortoise ownership.

Learn about the subtleties of caring for these amazing animals with a thorough nutritional guide, and investigate the different stages of their life cycle, from cute hatchlings to hardy adults. You can also acquire knowledge about environmental enrichment techniques that will keep your Horsefield Tortoise happy and healthy.

You'll gain a greater understanding of these kind animals as we cover subjects like bonding, socialization, and substrate selection. Whether you're a seasoned reptile enthusiast or a beginner, this guide will hopefully provide you the information and abilities you need to give your Horsefield Tortoise the best care possible.

Accompany us on this educational expedition to discover the techniques for cultivating a robust and long-lasting relationship with your Horsefield Tortoise.

Chapter 1

Care for Horsefield Tortoises: Crucial Advice for a Contented Pet

Scientifically named as Agrionemys horsfieldii or Testudo horsfieldii, Horsefield Tortoises have gained popularity as pets because of their small size, endearing personalities, and hardy nature. These reptiles are native to arid regions of Central Asia, and they require special care to ensure that they thrive in captivity. In this extensive guide, we will go over the most important things you should do to ensure the happiness and well-being of your Horsefield Tortoise.

1. Recognizing the Natural Environment:
Due to their adaptation to dry and semi-arid habitats, like the steppes of Central Asia, it is imperative that you recreate this natural habitat for Horsefield Tortoises.

You can accomplish this by utilizing a terrarium or outdoor enclosure with a substrate that resembles the sandy soil that they would find in their native locations.

2. Lighting and Temperature:

Horsefield tortoises require proper temperature and lighting maintenance. A basking spot should be kept between 90 and 95°F (32 and 35°C), and a cooler area should be kept between 75 and 85°F (24 and 29°C). Full-spectrum UVB lighting is also necessary for the metabolism of calcium in the tortoises and for their general well-being.

3. Correct Setup for Enclosures:

Provide a safe and cozy habitat for your tortoise. It's important to have a large enclosure with hiding places and a shallow water dish for bathing. You can also add natural elements to their habitat, like as logs and pebbles, to help stimulate their minds.

4. Proper Nutrition:

Your Horsefield Tortoise needs a diet that is well-balanced; offer leafy greens like kale, collard greens, and dandelion greens; supplement with calcium and vitamin D3 to help prevent nutritional deficiencies.

5. Stay Hydrated:

Make sure there's always clean, fresh water available. You should also have a shallow dish of water available for soaking, which aids in optimal shell health and hydration.

6. Frequent visits to the vet:

Plan routine checkups with a veterinarian who specializes in reptiles to keep an eye on your tortoise's health. Early detection of any problems allows for prompt intervention and treatment.

7. Managing and Socialization:

Horsefield tortoises are not as social as other pets, but they still need to be gently socialized. Handle them carefully so they become used to you being around, but don't handle them too much because they require time alone.

8. Selection of Substrates:

For digging and nesting behaviors, a mixture of topsoil, coconut coir, and cypress mulch can be used to create a ground that closely resembles their natural habitat.

9. Enhancement of the Environment:

Providing your tortoise with novel experiences and a varied environment will help to keep him or her happy and healthy. Some examples of enriching environments are hiding places, climbing structures, and toys to examine.

10. A Look at Breeding Considerations

Research and comprehend the duties associated with breeding Horsefield Tortoises. Breeding should be done with great care for the welfare of the adults as well as any offspring.

11. Recognizing and Dealing with Health Concerns:

Keep a close eye on your tortoise's look and behavior. Abnormalities in its shell, changes in appetite, or lethargy are all indicators of sickness. If you detect any worrying symptoms, get in touch with your veterinarian right once.

12. Hibernation and Seasonal Variations:

As hibernation replicates their natural behavior and can be helpful when done appropriately, it's important to research and follow right methods if you wish to allow your Horsefield Tortoise to hibernate, as this is a behavior known to occur in the wild.

13. Conscientious Ownership:

Taking care of a Horsefield Tortoise requires a long-term commitment; learn about their unique requirements and be ready for a lifetime that can reach several decades. Good ownership practices help these fascinating reptiles live longer and healthier lives.

14. Legal Points to Remember:

Before purchasing a Horsefield Tortoise, find out about any legal requirements or restrictions that may apply in your area. Certain regions may have laws governing the ownership of specific species of reptiles.

15. Maintaining Records:

Keep a journal of your tortoise's health, nutrition, and behavior. This will come in handy for you to return to later and for accurately informing a veterinarian should your tortoise become unwell.

To sum up, giving your Horsefield Tortoise the necessary care requires a combination of expertise, dedication, and sincere love for these fascinating reptiles. By comprehending and putting these suggestions into practice, you are not only guaranteeing your pet's physical health but also adding to their overall contentment and well-being. Keep in mind that every tortoise is an individual, and you can customize their care to meet their individual needs by observing their behaviors and preferences. Savor the adventure of friendship with your Horsefield Tortoise, and bask in the delight of taking care of such an amazing and resilient species.

Chapter 2

Comprehending Horsefield Tortoise Behavior

Exploring the world of Horsefield Tortoises involves more than just offering a suitable habitat; it also involves understanding their complex behaviors. As individual reptiles, Horsefield Tortoises display behaviors that carry significant messages about their overall health, well-being, and general condition. In this investigation, we will dissect the different facets of their behavior, offering insights into their habits, communication, and responses to their surroundings.

1. Foraging and Exploration:
Even though they move slowly, Horsefield Tortoises are naturally curious about their surroundings. They use their powerful legs to walk around their habitat in search of food and good places to bask. This activity lets

them interact with their environment and stimulates their minds in addition to providing them with sustenance.

2. Heat-regulating agents and basking

Horsefield Tortoises actively search out basking spots in their enclosure to absorb heat, which helps with digestion and maintains their metabolic processes. Horsefield Tortoises bask because they thermoregulate by exposing themselves to sunlight, and basking behavior can provide insights into their general well-being.

3. Excavation and Tunneling:

Digging and burrowing are natural behaviors for Horsefield Tortoises, particularly when it comes to nesting or hiding. By providing enough substrate in their cage, you may encourage this habit in them and help them fulfill their instincts while also improving their overall wellbeing.

4. Social Conduct:

Horsefield tortoises are not naturally gregarious animals, but they may exhibit some territorial tendencies. If kept in a group, they might establish dominance or protect their territory, so it's vital to keep an eye on things to avoid hostility and maintain a peaceful environment.

5. Body Language as a Form of Communication:

Horsefield Tortoises do not have many facial expressions, but they do communicate through body language. For example, head bobbing can indicate aggression or courtship. Recognizing these nonverbal cues can help create a setting that reduces stress and fosters a sense of security for the tortoise.

6. Retreating and Hiding:

As a natural defense strategy to assist them avoid potential attacks, Horsefield Tortoises frequently retreat into their shells when they feel anxious or threatened. Establishing hiding spaces within their cage is crucial to providing them with a sense of comfort.

7. Engaging with Items:

The tortoise's curiosity can be piqued by adding pebbles, logs, or even toys to their surroundings. They can push things around, engage with them, or use them as climbing frames. Having a variety of materials available will improve their surroundings and keep their minds active.

8. Reaction to Environmental Shifts:

Gradual introductions to changes can help decrease stress. Horsefield tortoises are creatures of habit, and unexpected changes in their surroundings can generate stress. Whether it's a new enclosure setup or a change in temperature, study their reactions to ensure they adapt easily.

9. Behavior During Mating and Courtship:
Male Horsefield Tortoises may exhibit head bobbing or circling during mating behavior. Knowledge of these behaviors is important for anyone thinking about breeding these tortoises, since it sheds light on the reproductive cycle.

10. Variations in Activity Levels and Seasons:
Like many other reptiles, horsefield tortoises have seasonal behavioral variations. They may become less active during the colder months, and knowing these variations can aid with care routine adjustments, such as feeding and lighting.

11. Cognition and Problem-Solving Skills:

Horsefield tortoises aren't usually thought of as very smart, but they can solve problems. They can find their way around, hide, and even identify their owners. They can also be made to show off their cognitive skills by giving them puzzles or challenges to solve.

12. Indices of Health:

Any abrupt changes in the tortoise's activity level, eating habits, or general behavior should be closely observed. Frequent observations enable the early detection of potential health issues, hence facilitating appropriate veterinarian care. Behavioral changes can be an indicator of the tortoise's health.

13. Hibernation and Sleep Patterns:

Horsefield tortoises have unique sleep habits; they frequently spend more of the day active and the night sleeping. If your tortoise is thinking about hibernating, it's important to know the natural indications and create the right environment to help it emulate its natural behavior.

14. Communication with People:

Establishing a link with your tortoise through gentle interactions and positive experiences can establish a sense of trust and reduce stress. Although Horsefield tortoises are not known for significant engagement, they can become acclimated to human presence and handling.

15. Extended Life Expectancy and Shifts in Behavior:

Given the lengthy lifespan of Horsefield tortoises, it is possible for their behavior to alter with age. It is important to recognize these changes in behavior and modify care to provide your tortoise with a happy and healthy existence.

To sum up, learning about the behavior of Horsefield Tortoises is an intriguing adventure that calls for perseverance, astute observation, and a respect for the nuances of their movements. By exploring their world and learning the meaning behind each behavior, you improve their quality of life and develop a closer bond with these amazing reptiles. The more we know about their behaviors, the more capable we are of creating an environment that satisfies their mental, physical, and emotional needs, resulting in a harmonious and rewarding relationship between the caretaker and the tortoise.

Chapter 3

Equine Enclosures: Crafting the Ideal Environment

Starting the process of building a Horsefield Tortoises dream home is a fulfilling undertaking that entails more than just giving them a place to live. It entails learning about their innate desires, instincts, and the delicate equilibrium needed to maintain their physical and mental health. In this extensive guide, we will go over the key components and factors to take into account when building a Horsefield Tortoises dream home.

1. Dimensions of the Enclosure:
For an adult Horsefield Tortoise, an enclosure should be at least 4 feet by 8 feet, but larger is always preferable. If the climate permits, consider outdoor enclosures, which allow for more space and exposure to natural sunlight. The size of the enclosure is a critical factor in

creating a suitable habitat for Horsefield Tortoises. These reptiles require ample space to move, explore, and express their natural behaviors.

2. Selection of Substrates:

A mixture of topsoil, coconut coir, and cypress mulch creates a substrate that encourages digging and burrowing behaviors; this mix also helps to maintain optimal humidity levels and offers a comfortable surface for the tortoise. Selecting the appropriate substrate is crucial to replicating the natural habitat of the Horsefield Tortoise.

3. Gradient of Temperature:

Enclosures must have a temperature gradient in order for the tortoise to regulate its body temperature. The warmer side of the enclosure should be between 75 and 85 degrees Fahrenheit (24 and 29 degrees Celsius), and the basking area should be between 90 and 95 degrees

Fahrenheit (32 and 35 degrees Celsius) to allow for the best possible digestion and metabolic processes. Frequent temperature monitoring will guarantee a comfortable and healthy environment.

4. Lighting UVB:

For the purpose of promoting calcium metabolism and preventing metabolic bone disease, full-spectrum UVB lighting is crucial for Horsefield Tortoises. Place UVB lamps over the basking area so the tortoise can have access to this light source for a good portion of the day. Replace UVB bulbs on a regular basis to keep them functional.

5. Cover and Hide Outs:

Enough hiding places to relieve stress and encourage natural behaviors are important for Horsefield Tortoises, who enjoy having safe havens and hiding places to retreat to. These can be made from natural materials such as logs, pebbles, or commercially supplied hides.

6. Water Elements:

Since Horsefield Tortoises cannot swim, they benefit from having a shallow water dish for drinking and soaking. Make sure the dish is easily accessible and cleaned on a regular basis to keep it hygienic. It is important to keep an eye on their hydration levels, and having a designated space for soaking helps to prevent dehydration.

7. Climbing Elements:

The addition of climbing structures to the enclosure offers variation and encourages the tortoises' natural behaviors. Placement of flat rocks, logs, or specially designed climbing structures can stimulate exploration and physical activity, thereby improving the environment and offering opportunities for natural exercise.

8. Grazing land and edible plants:

Create a designated grazing area to stimulate foraging, which offers both mental stimulation and a variety diet. Plants that are safe for tortoises, such as hibiscus, dandelion, and other grasses, can be included in the enclosure to allow Horsefield Tortoises to engage in their natural grazing habits.

9. Refeeding Points:

Establish dedicated feeding stations within the enclosure to assist preserve cleanliness and allow for a more structured approach to delivering a well-balanced and varied diet, as well as to prevent food contamination and simplify the monitoring of the tortoise's diet.

10. Enhancement of the Environment:

You may prevent boredom in Horsefield Tortoises by adding different aspects to the enclosure, such as rotating and introducing new things, frequently rearranging the arrangement, or creating a digging area with loose substrate, to keep them cognitively active.

11. Design of Secure Enclosures:

Since Horsefield Tortoises are skilled diggers, it is imperative to make sure the enclosure is safe and escape-proof. You should also routinely check the enclosure for any possible weaknesses or wear in the fencing materials.

12. Planting Plants That Are Friendly to Tortoises:

Select and study plants that are safe and non-toxic for tortoises before planting them directly in the enclosure. This will enhance the enclosure's beauty while also offering more hiding places and the possibility of natural foraging.

13. Keeping Things Tidy:

For the Horsefield Tortoise to remain healthy, regular cleaning and care are essential. Spot-clean the substrate, clean the water dish on a regular basis, and remove any uneaten food as soon as possible. Perform a more comprehensive cleaning once a month to avoid the accumulation of bacteria and waste.

14. Keeping an eye on behavioral cues:

Understanding the needs of the tortoise requires paying close attention to how it behaves in its enclosure. If it frequently hides or avoids a particular area, there may be a need to make changes to its surroundings. By designing the enclosure in accordance with the tortoise's behavioral cues, you can create a stress-free living environment for it.

15. Seasonal Modifications:

If your Horsefield Tortoise is subject to natural variations in temperature and daylight hours, you may want to think about making seasonal modifications to the enclosure. For example, adding extra heating during the winter months or modifying the timing of the lighting simulates the natural habitat in which they live.

To sum up, creating the ideal home for Horsefield Tortoises requires a comprehensive strategy that takes into account their innate tendencies, physical needs, and psychological health. In order to provide children with a happy and meaningful existence, it is our responsibility as responsible caregivers to create an atmosphere that not only satisfies their fundamental requirements but also permits them to express their natural tendencies. You can establish a habitat that not only mimics the Horsefield Tortoises native environment but also lays the groundwork for an enduring and solid relationship between the caretaker and the tortoise by combining meticulous design, close attention to detail, and continuous observation.

Chapter 4

How to Feed Your Horsefield Tortoise: A Guide to Nutrition

A healthy diet is essential to the health and welfare of Horsefield Tortoises. These reptiles require a broad and well-balanced diet to flourish because they are herbivores with particular dietary needs. We will examine the dietary requirements of Horsefield Tortoises in this extensive nutritional guide, including advice on appropriate meals, feeding schedules, and necessary supplements to ensure a healthy and happy life for your beloved companion.

1. The Wild Diet Naturally:
Comprehending the dietary requirements of Horsefield Tortoises in captivity requires an understanding of their natural diet in the wild. These tortoises graze on a

variety of grasses, leafy greens, and occasionally flowers in their natural habitat of Central Asia. Their vigor and health are founded on this plant-based diet.

2. Greens with leaves:

Leafy greens are the main component of a Horsefield Tortoise's diet since they are a good source of fiber, vitamins, and minerals. Incorporate a range of greens, including turnip, mustard, collard, and dandelion greens. These should make up the majority of the meals they eat each day.

3. Hay and Grass:

Horsefield Tortoises need high-fiber hay and grasses for their digestive systems to stay healthy. Bermuda grass hay and Timothy hay are great options. These fibrous substances help to keep the gut motile and guard against problems like impaction.

4. Edible Flowers and Weeds:

The varied food that Horsefield Tortoises enjoy includes edible weeds and flowers. Dandelion, clover, hibiscus blossoms, and nasturtiums are safe choices. These additions encourage natural foraging habits in addition to offering a range of nutrients.

5. Produce:

Include a little quantity of veggies in the diet to improve the diversity of nutrients. Cucumbers, zucchini, bell peppers, and carrots are safe options. On the other hand, stay away from oxalates-rich veggies like kale and spinach since too much of them might cause calcium-binding.

6. Fruits:

Due to their greater sugar content, fruits including strawberries, melon, and papaya should only be served occasionally as treats. Fruits should be used as a supplement rather than a main source of nutrition because too much sugar can lead to obesity and other health problems.

7. Supplementing with Calcium:

For the growth and upkeep of a Horsefield Tortoise's shell as well as the general health of its bones, calcium is essential. Give leafy greens and veggies a dusting of vitamin D3-rich calcium supplements, or give tortoises a cuttlebone to chew on. To avoid metabolic problems, make sure your calcium-to-phosphorus ratio is appropriate.

8. Control of Phosphorus:

Keep the ratio of phosphorus to calcium in your diet in check. The absorption of calcium might be hampered by an excess of phosphorus. Restrict your intake of foods high in phosphorus, like some nuts and seeds, and concentrate on giving your child a diet high in calcium and variety.

9. Avoid Diets High in Protein:

High protein diets are not necessary for Horsefield Tortoises, and consuming too much protein can cause kidney problems. Reduce your intake of foods heavy in protein, such as meat and vegetables, and place more emphasis on getting the right amount of fiber, vitamins, and minerals.

10. Feeding Pattern:

To control the amount of nutrients consumed, set up a regular eating plan. While younger tortoises can need more frequent feeds, adult Horsefield tortoises usually benefit from daily feedings. Keep an eye on their weight and modify the portion sizes appropriately.

11. Drinking plenty of water

For the Horsefield Tortoises to be healthy, they must drink enough water. Give them a small water dish to drink from while they get some moisture from their food. Additionally, a few times a week, giving your tortoise a soak in lukewarm water encourages hydration and guards against dehydration.

12. Seasonal Differences:

Recognize that the food of your tortoise varies with the seasons. They are more active in the summer months, so you might notice an increase in hunger. Adapt feeding schedules and quantities accordingly. Their metabolism may slow down in the winter, so you may need to feed them less during those months.

13. Steer clear of toxic plants:

Learn about the plants that are hazardous to tortoises and make sure they don't eat them. Common garden plants that might be dangerous include ivy, rhubarb, and several flowers. Prior to adding additional plants to the cage, do your homework.

14. Track Your Weight and Physical Condition:

Keep a regular eye on the weight and physical condition of your Horsefield Tortoise. A healthy tortoise should resemble a little domed object and have a well-rounded shell. See a veterinarian right away if you observe weight loss or changes in the shape of your shell.

15. Rotational Food Scheduling:

Put into practice a rotating diet plan to add diversity and guard against nutritional deficits. Alternate between various leafy greens, veggies, and edible weeds to guarantee a wide range of nutrients. This strategy resembles how they would naturally forage in the wild.

16. Breeding Factors:

When breeding Horsefield Tortoises, it is especially important to consider the dietary requirements of gravid (or pregnant) females. To encourage the formation of eggs and effective reproduction, supplement with calcium and make dietary adjustments.

17. Veterinary Examinations:

To determine your tortoise's general health and make sure their dietary demands are satisfied, routine veterinary examinations are essential. A veterinarian who specializes in reptiles can offer advice on diet modifications depending on the age, health, and special needs of the tortoise.

18. Updates and Educational Resources:

Keep up with developments in the diet and care of reptiles. A better understanding of the health of Horsefield tortoises and new study findings could result in food recommendations being updated. Keep up to date by participating in reliable forums, literature, and veterinary resources for reptiles.

To sum up, feeding your Horsefield Tortoise is a complex part of responsible ownership that calls for knowledge, commitment, and close attention to detail. You enhance their longevity, energy, and general well-being by feeding them a varied, well-balanced diet that closely resembles their natural foraging practices. Adjust their diet to suit their demands, pay attention to their tastes, and take pleasure in knowing that you are giving your Horsefield Tortoise the best possible nourishment for a happy and healthy animal.

Chapter 5

Horsefield Tortoises Common Health Problems and How to Treat Them

A strong awareness of potential health issues and attentive care are necessary to ensure the well-being of a Horsefield Tortoise. These tortoises are resilient reptiles, yet they are nevertheless prone to various health issues, some of which can be avoided or lessened with good care. This thorough guide will examine frequent health problems that Horsefield Tortoises face and offer information on their causes, symptoms, and practical preventative and treatment methods.

1. infected respiratory systems:
   - Causes: Inadequate temperature or high humidity are two common environmental factors associated with respiratory illnesses in Horsefield

Tortoises. Respiratory problems can also be caused by bacterial or viral causes.

- Symptoms: Wheezing, nasal discharge, open-mouth breathing, and lethargic behavior are indications of respiratory infections.

- Prevention and Treatment: Keep your tortoise at the right temperature, make sure it has enough ventilation, and keep it out of the draft. If you experience any respiratory issues, get veterinarian help right away. Both supportive care and antibiotics may be used in treatment.

2. Rotting Shells:

- Causes: When a tortoise's shell is regularly exposed to a moist environment, it can become infected with bacteria or fungi, which is what causes shell rot. Injuries, insufficient substrate, and poor hygiene can all cause shell rot.

- Symptoms: Shell rot may be indicated by discoloration, mushy patches on the shell, or an unpleasant smell. When the infection is severe, the bone may become infected.

- Prevention and Treatment: Make sure your tortoise has access to a dry basking area and a clean, dry substrate. Keep the enclosure clean on a regular basis to maintain good hygiene. For a precise diagnosis and suitable treatment—which can include taking antibiotics or antifungal drugs—seek veterinarian care.

3. MBD, or metabolic bone disease:

- Causes: A deficit in calcium and vitamin D3 leads to metabolic bone disease. MBD might arise as a result of inadequate UVB exposure, an unbalanced diet, or incorrect supplementation.

- Symptoms: MBD symptoms include limb tremors, mobility issues, and softening or abnormalities in the shell.

- Prevention and treatment options include giving appropriate UVB exposure, supplying calcium supplements, and providing a well-balanced, high-calcium diet. Veterinarian intervention, including as supportive care and calcium injections, may be required in extreme situations.

4. Parasitic Diseases:

- Causes: Horsefield tortoises may be impacted by internal parasites including worms and protozoa. Parasitic diseases can result from contaminated food, water, or contact with infected animals.

- Symptoms: Changes in the appearance of feces, weight loss, diarrhea, and lethargy may all be signs of parasite infections.

- Prevention and Treatment: Provide clean food and water, maintain a clean environment, and practice excellent hygiene. Effective detection and treatment of parasite illnesses are facilitated by routine veterinary inspections of the feces.

5. Eye Issues:

- Causes: Injuries, illnesses, or underlying medical diseases can all lead to eye problems in Horsefield Tortoises. Stressful surroundings or insufficient humidity could also be factors.
- Symptoms: Eye issues might manifest as discharge, excessive tearing, or swollen, red, or hazy eyes.
- Prevention and Treatment: Maintain appropriate humidity levels, keep the surroundings tidy, and keep a watch out for symptoms of eye discomfort. For a proper diagnosis and suitable treatment—

which can include antibiotic eye drops—consult a veterinarian.

6. Binding of Eggs:

- Causes: A variety of causes, including inadequate calcium levels, dehydration, or unsuitable nesting conditions, might make it difficult for female Horsefield tortoises to lay eggs.
- Symptoms: Egg binding may be indicated by lethargy, agitation, straining, or digging without laying eggs.
- Prevention and Treatment: Provide a dignable substrate and a suitable nesting spot. Make sure you eat a calcium-rich diet and stay properly hydrated. Egg-binding can be a potentially fatal illness, so get veterinary help right once if you suspect it.

7. Dehydration:

- Causes: Inadequate water availability, arid weather, or illnesses that impair hydration can all lead to dehydration in tortoises.

- Symptoms: Dehydration manifests as sunken eyes, fatigue, dry skin, and irregular urine patterns.

- Prevention and Treatment: Offer a drinking dish with shallow water and consistent submersion. Make that the humidity inside the cage is at the right level. It could be necessary for veterinarians to provide fluid treatment in cases of severe dehydration.

8. Impact:

- Causes: When a tortoise eats substrate or other indigestible materials, the digestive tube becomes blocked, which causes impaction. Inappropriate food consumption or limited access to a healthy diet may be the cause of this.

- Symptoms: Impaction may manifest as a bloated or enlarged abdomen, decreased appetite, and fatigue.

- Prevention and Treatment: Provide a good substrate to aid with digestion, and keep an eye on the tortoise's food to make sure it doesn't include anything that can't be broken down. See a veterinarian for diagnosis and treatment; in extreme cases, surgery and/or hydration may be necessary.

9. Overgrown Nails and Beak:

- Causes: If a tortoise's diet does not provide enough abrasion or if it does not have access to surfaces that naturally erode these structures, it may develop overgrown beaks and nails.

- Symptoms: Overgrowth manifests as difficulty eating, excessive beak growth, or untrimmed, curled nails.

- Prevention and Treatment: Offer naturally occurring items that abrasively surface areas, like rocks. Keep an eye on growing beaks and claws and regularly trim them, or seek veterinarian care for expert attention.

10. Spot rot (stomatitis):

- Causes: Injuries, an unbalanced diet, or bacterial or fungal infections can all lead to stomatitis, an inflammation of the oral tissues.
- Symptoms: Stomatitis may manifest as swelling, drainage, or trouble chewing.
- Prevention and Treatment: Keep your mouth clean, feed it a healthy diet, and keep an eye out for any irregularities. For a precise diagnosis and treatment—which can include taking antibiotics or antifungal drugs—see a veterinarian.

11. Problems with Hibernation:

- Causes: If the tortoise is not well prepared for hibernation or if the settings are not regulated, hibernation-related health issues may arise.
- Symptoms: Lethargy, trouble moving, and inability to awaken from hibernation are indications of problems associated to hibernation.
- Prevention and Treatment: Make sure you are properly prepared for any scheduled hibernation, which should include gradual cooling and water. Keep a watchful eye on hibernating tortoises and seek prompt medical care if problems develop.

12. Modifications in Behavior:
- Causes: Behavioral changes like hiding more often, becoming less active, or changing how you eat could be signs of underlying health problems.
- Symptoms: Though they can vary greatly, behavioral changes frequently indicate stress, illness, or discomfort.

- Prevention and Treatment: Pay close attention to your tortoise's behavior and take quick action if it changes. For a complete health evaluation, consult a veterinarian and keep your pet in a constant and appropriate environment.

13. Injuries and Trauma:
- Causes: Accidents, collisions, and interactions with other animals can all lead to trauma and injury. Environmental risks or improper handling could also be factors.
- Symptoms: After an incident visible wounds, limping, or behavioral changes may be signs of trauma.
- Prevention and Treatment: Provide a secure atmosphere, stay away from handling that could injure or stress people, and keep an eye out for any dangers. See a veterinarian for a comprehensive checkup and the necessary care.

14. Dystocia, or trouble laying eggs:

- Causes: Dystocia is a condition where a female tortoise has trouble producing eggs. This is frequently brought on by issues with egg size, defects, or a lack of suitable nesting sites.

- Symptoms: Dystocia may be indicated by signs of distress, lethargy, and recurrent digging without egg laying.

- Prevention and Treatment: Make sure there's a good place for the nest, make sure the right amount of calcium is consumed, and keep an eye out for nesting activity. Seek prompt veterinarian assistance for assessment and possible intervention if dystocia is suspected.

15. heightened levels of stress

- Causes: A number of things, such as poor handling, unsuitable environmental circumstances, or environmental changes, can lead to stress.

- Symptoms: Common indicators of stress include changing activity levels, decreased appetite, and behavioral changes.
- Prevention and Treatment: Keep the surroundings consistent, try to avoid causing too much disturbance, and handle the tortoise sparingly and carefully. To avoid extended stress, quickly identify and deal with stressors.

In conclusion, good husbandry, careful attention, and early detection and resolution of possible problems are all critical to the health of a horsefield tortoise. A balanced diet, regular veterinary exams, and a comfortable surroundings all contribute to their longevity and general well-being. Knowing the common health problems mentioned above and taking preventative action will help you make sure your

Horsefield Tortoise survives in captivity and brings you years of happiness and companionship. Always seek the advice of a veterinarian who is knowledgeable with reptiles for expert advice catered to the unique requirements of your tortoise.

Chapter 6

Horsefield Tortoises: A Life Cycle from Hatching to Adulthood

Agrionemys horsfieldii, also known as Testudo horsfieldii, are fascinating animals whose lives are characterized by a fascinating trip through several stages of growth and development. These reptiles change in size, behavior, and preferred habitat as they mature from hatchlings to adults. It is essential to comprehend each step in order to give the best care possible and guarantee the wellbeing of these tenacious and charming animals. We will examine the many stages of the Horsefield Tortoise lifecycle in this thorough guide, illuminating their special traits, requirements, and crucial factors at each turn.

1. Incubation of Eggs:

The process of egg incubation marks the beginning of a Horsefield Tortoise's cycle. Usually, female tortoises select a specific nesting place where they lay a clutch of eggs. The eggs are buried in the substrate to shield them from the environment and predators. The eggs are typically spherical and leathery. Although it might vary, the incubation time typically lasts between 60 and 90 days.

2. Hatching:

The little tortoises emerge from their eggs when the incubation period is over. The delicate, malleable shell that hatchlings emerge with eventually hardens over time. Hatching is an important stage in the life cycle, and the hatchlings have the natural ability to excavate a passage out of the nest.

3. Dimensions and Look:

The Hatchling Horsefield Tortoises are extraordinarily small; their average length is between 1.5 and 2 inches. When compared to the vivid and distinctive patterns that will emerge as they mature, their shells are quite pale. These juvenile tortoises have charming characteristics, such as big eyes and a charming hint of vulnerability.

4. Action and Investigation:

Inquisitive by nature, hatchlings show a desire to investigate their environment. They exhibit a habit that balances intervals of rest and exploration. In order to nurture their natural tendencies and encourage healthy development throughout this stage, it is essential to provide a safe and stimulating environment.

5. Nutrition and Feeding:

Horsfield Hatchling Tortoises have particular dietary needs. Their growth and shell formation depend on a diet high in leafy greens, finely cut vegetables, and a calcium supplement. A healthy and happy tortoise begins with ensuring adequate nutrition during this formative period.

6. Configuration of the Habitat:

Hatchling enclosures should be scaled suitably to the size of the animals inside. It is necessary to have a modest, safe area with a shallow water dish, hiding places, and a temperature gradient for basking. The soft substrate and the entire setting should replicate the comfort and safety they found in the nest.

7. UV-B Radiation:

For hatchling Horsefield Tortoises, it is essential to provide sufficient UVB lighting. For appropriate calcium metabolism and the development of a robust and healthy shell, exposure to UVB radiation is necessary. It is important to place UVB lamps so that the hatchlings get the UVB rays they require.

8. Rate of Growth:

A Horsefield Tortoise's growth rate is comparatively fast during its early life. Healthy growth is facilitated by a good habitat, availability to UVB lighting, and proper diet. Keeping an eye on the tortoise's weight and size might reveal important information about their general health.

9. Making the Move to a Bigger Enclosure:

The Horsefield Tortoise hatchling eventually outgrows its original container as it matures. Moving to a broader area makes it possible to add additional enriching aspects, enhance activity, and carry on with exploration. This change takes into account their increasing numbers and levels of activity.

10. Teenage years:

The teenage stage is characterized by ongoing development and growth. At this stage, Horsefield Tortoises are more gregarious and may exhibit mature behaviors. Adolescents, who are not quite adults, show greater independence and could start becoming pickier about what they eat.

11. Nutritional Modifications:

The food of adolescent Horsefield tortoises may need to be modified to meet their evolving nutritional requirements. A diversified and well-balanced diet, with a base of leafy greens and vegetables, is beneficial to their general health. Supplementing with calcium is still crucial for the development of shells.

12. Shell Creation:

The Horsefield Tortoise's shell continues to harden and take on its distinctive markings during adolescence. The shell's general appearance brightens and the bony plates (scutes) become more pronounced. A healthy diet and exposure to UVB sun are essential components of this continuous process.

13. Social Structure:

Horsefield tortoises may exhibit specific social behaviors as they approach maturity. Even if people are not naturally sociable beings, interactions can nonetheless happen, particularly if they live together. In order to maintain a peaceful home and stop hostility, it is essential to keep an eye on these dynamics.

14. Mature Reproduction:

The sexual maturity of Horsefield Tortoises usually occurs between the ages of 5 and 8 years, although this might vary based on genetics, habitat, and food. Certain actions, such head bobbing, circling, or trying to mount females, are displayed by males during courtship. Females may act in a nesting manner.

15. Nest Building and Procreation:

Once sexually mature, female Horsefield tortoises may engage in nest-related activities such as digging and looking for suitable locations. For females to lay eggs, a specific place with suitable substrate must be provided. Understanding the reproductive cycle is crucial for appropriate breeding practices for those who are thinking about starting a family.

16. Laying of Eggs and Incubation:

In meticulously constructed nests, females bury their eggs in the substrate. These eggs incubate for 60 to 90 days, which is similar to the time it takes for the first hatchlings. For this reproductive stage to be successful, an environment that is favorable for nesting and egg laying must be created.

17. Limited Parental Care:

Horsefield tortoises do not display a great deal of parental care, in contrast to certain other species. The females might not be directly involved in raising the young once the eggs are laid. Hatchling survival and development are primarily dependent on the surrounding environment and the appropriate care given by the tortoise keeper.

18. Sustained Expansion and Durability:

With the right care, Horsefield Tortoises can live for several decades and continue to grow until adulthood. The shell of an adult tortoise is stronger, more fully developed, and has unique colors and patterns. Because of their extended lifespan, these reptiles require long-term care from their caregivers.

19. Continued Care in Later Life:

Continuing to care for adult Horsefield Tortoises entails keeping their surroundings steady and stimulating. Providing a balanced diet, making sure UVB exposure is appropriate, and keeping an eye out for any indications of health problems are all part of this. Regular veterinary examinations are becoming more and more crucial to guaranteeing their long-term health.

20. Environmental Modifications:

Certain environmental adaptations may be seen in adult Horsefield tortoises. They learn how to find their way around their enclosures, make use of their basking places, and seek cover when necessary. By getting to know their preferences and behaviors, caregivers can design a space that honors their innate desires.

21. Lifespan and Aging:

Horsefield tortoises may become less active and more sedentary as they get older. It is essential to comprehend aging in order to modify their care regimens. These tortoises have a remarkable and enduring longevity that allows them to survive well into their 50s or even beyond with the right care.

In summary, the life cycle of Horsefield Tortoises is an amazing voyage that includes phases of development, maturation, and adaptability. These reptiles enthrall aficionados with their distinct traits and behaviors, from the small, delicate hatchlings emerging from their eggs to the robust and long-living adults. The general well-being of these hardy and intriguing animals depends on providing careful attention at every stage, comprehending their changing needs, and modifying the surroundings appropriately. For those who take care of them, appreciating each stage of their life cycle guarantees a happy and long-lasting bond with these adorable reptile friends.

Chapter 7

Extension of the Environment for an Intensely Motivated Horsefield Tortoise

Horsefield tortoises (Agrionemys horsfieldii or Testudo horsfieldii) require a dynamic and exciting environment to be healthy. These hardy reptiles are indigenous to dry areas of Central Asia, and careful habitat design can enhance their natural habits in captivity. This thorough guide will cover a variety of techniques and components to improve Horsefield Tortoises' habitat enrichment, encouraging movement, cerebral stimulation, and general well-being.

1. Layout and Size of Enclosures:
Enclosures of a suitable size are the cornerstone of environmental enrichment. Despite their diminutive size, Horsefield Tortoises like a large living space. For an

adult tortoise, a minimum enclosure size of 4 feet by 8 feet is advised, providing plenty of room for the tortoise to walk around, explore, and exhibit its natural habits.

2. Variety of Substances:

For Horsefield Tortoises, providing a varied substrate improves their sensory experience. An appropriate foundation for digging and burrowing is made of a combination of dirt, coconut coir, and cypress mulch. This diversity of substrates encourages natural activities and adds to the overall enrichment of their living environment.

3. Climbing Frameworks:

The popular belief is that tortoises are only found on the ground, but Horsefield Tortoises also like to climb and explore higher ground. Add logs, flat pebbles, or specially designed climbing frames to promote vertical movement. These buildings not only improve their surroundings but also offer chances for physical activity.

4. Hiding Places & Cover:

In order to lessen stress and enable tortoises to withdraw when necessary, hiding places and shelters must be established. Logs, boulders, and commercially available hides are examples of natural items that add to the enclosure's enrichment. These hiding places provide safety and resemble the natural behavior of tortoises in the wild.

5. Excavation Region:

Given that Horsefield Tortoises love to dig, creating a special location for them to do so enriches their surroundings greatly. They can dig in a sand pit or other area of loose substrate as it suits their natural digging habits. This stimulating activity encourages both mental and physical exercise.

6. Grazing land and edible plants:

Adding edible plants to the cage encourages natural grazing behaviors while also improving its aesthetic appeal. Choose plants that are safe for tortoises, like grasses, hibiscus, and dandelion. Establish a special grazing area to promote foraging, which will support a varied diet and mental stimulation.

7. Warming Spots with Variations in Temperature:

For Horsefield Tortoises to maintain thermoregulation, basking places with temperature gradients must be established. Temperatures between 90 and 95°F (32 and 35°C) are ideal for basking because they promote healthy metabolic and digestive functions. Make sure the side that is cooler has a temperature range of 75–85°F (24–29°C) so the tortoise can comfortably control its body temperature.

8. Water-related Features:

Although they cannot swim, Horsefield Tortoises gain from having a shallow water dish for bathing and drinking. Including a water feature gives the enclosure more movement, like a shallow dish or a low-sided container. The water is suitable for tortoises to sip from and soak in, which is beneficial to their general health.

9. Placement of Rotational Objects:

Rotate and add new elements to the enclosure on a regular basis to enhance the habitat. This can involve adding new objects, relocating pebbles, or altering the arrangement of hiding places. Rotational object placement keeps things interesting and piques the tortoise's interest.

10. Natural Tunnels and Hides:

Emulating Horsefield Tortoises' innate burrowing behavior makes them feel more secure and gives them a richer experience. To promote digging and exploration, incorporate natural hides and burrows by placing rocks, logs, or soil mounds in strategic locations.

11. UVB Radiation with Exposure to Natural Sunlight:

Horsefield tortoises require access to natural sunshine or the use of UVB lamps. It is essential for calcium metabolism and general health to be exposed to UVB rays. Place UVB lamps above basking regions to guarantee that tortoises have access to this essential source of light. Make sure they have access to both sunny and shaded places if they are housed outside.

12. Interactive Kitchen Sinks:

Establish interactive feeding stations to turn mealtimes into learning experiences. Disperse food items throughout the enclosure to encourage the tortoise to explore and get food on its own. This satisfies their innate desires and keeps them from becoming bored with regular feeding schedules.

13. Perceptual Cues:

Use sensory stimulation to appeal to a variety of senses. Because tortoises are visual and tactile learners, it is interesting to introduce a range of textures, colors, and forms to their surroundings. Various kinds of rocks, textured surfaces, or even safe objects with varied surfaces can be examples of this.

14. Introducing Toys Safe for Tortoises:

Horsefield tortoises can be curious about some objects, even if they are not as playful as some other pets. Toys safe for tortoises, like textured objects or firm rubber balls, should be introduced. These objects offer chances for investigation and engagement.

15. Feeding Obstacles and Puzzle Playthings:

To improve the feeding experience, including games and puzzles. Food can be hidden in substrate or placed inside puzzle feeders to encourage the tortoise to solve problems and engage in more active foraging.

16. Seasonal Modifications:

Take into account modifying the enclosure according to the tortoise's seasonal needs. Since tortoises tend to be more active during the warmer months, giving them extra enrichment—like new objects or rearranged features—will keep them interested. Changes in illumination and temperature are examples of adjustments.

17. Surfaces with Mirrors or Reflections:

To add visual interest, add a mirror or other reflective surface to the enclosure. In order to bring some novelty into their surroundings, tortoises may show signs of curiosity or interest in their own reflections. Keep an eye on their reaction to make sure it doesn't stress them out.

18. Water Features That Interact:

Use a shallow pool or tray to create an interactive water feature where the tortoise can relax and soak. Enhancing this space with floating items or edible water plants can increase its appeal. Some tortoises get a unique sensory experience when they splash around or investigate water features.

19. Observation and Reactivity:

Keep a close eye on the tortoise's actions and how well it responds to enrichment. If certain elements or objects seem to catch their attention, think about including more of the same in the future. Customized enrichment of the environment is made possible by being sensitive to their preferences.

20. Aging-Related Environmental Changes:

Understand that as they age, Horsefield Tortoises may have different needs and preferences in terms of their surroundings. Providing opportunities for greater mobility, exploration, and environmental complexity becomes more crucial as they get bigger. Changes should be made to account for their evolving needs and behaviors.

21. Engaging and Observing:

Even though Horsefield Tortoises might not be as socially inclined as other pets, spending time watching them interact with one another helps to establish a stronger bond. Honor each one of their unique personalities and inclinations, and enjoy watching them as they are in their enhanced surroundings.

22. Frequent Audits for Enrichment:

Conduct regular enrichment audits to evaluate the features and enclosure layout's efficacy. Determine whether the tortoise interacts with different objects and make necessary modifications to keep the habitat interesting. A dynamic and ever-changing feature of their captivity habitat ought to be enrichment.

23. A Look at Holistic Health:

Horsefield Tortoises' general health and well-being are linked to environmental enrichment. Creating an environment that is stimulating encourages behavioral satisfaction, mental and physical exercise, and a more comprehensive approach to their care.

In summary, the establishment of an enhanced living space for Horsefield Tortoises necessitates a deliberate fusion of habitat planning, behavioral analysis, and continuous observation. With a variety of components that support their innate instincts and behaviors, caregivers can guarantee that these hardy reptiles have happy, stimulating lives in captivity. Creating an environment that is dynamic, diverse, and ever-changing while supporting their physical and mental well-being and enabling them to express their individual personalities and behaviors is crucial.

Chapter 8

Selecting the Appropriate Substratum for Your Equine Companion

Selecting the appropriate substrate for your Horsefield Tortoise (Agrionemys horsfieldii or Testudo horsfieldii) is a crucial aspect of creating a healthy and comfortable environment. The substrate serves various purposes, including facilitating natural behaviors, aiding in thermoregulation, and contributing to the overall well-being of your tortoise. In this comprehensive guide, we will explore the factors to consider when choosing a substrate for your Horsefield Tortoise and delve into the different options available, ensuring you can make an informed decision that aligns with the needs of your reptilian companion.

1. Natural Habitat Considerations:

Understanding the natural habitat of Horsefield Tortoises is the foundation for selecting an appropriate substrate. In the wild, these tortoises inhabit arid regions with sandy or loamy soil. Mimicking these conditions in captivity helps create a familiar and comfortable environment for your tortoise.

2. Adequate Drainage:

One of the primary functions of substrate is to provide adequate drainage. Horsefield Tortoises are prone to respiratory issues if kept in overly damp conditions. Choose a substrate that allows water to drain effectively, preventing the accumulation of moisture in the enclosure.

3. Soft and Burrowable:

Horsefield Tortoises are avid diggers and burrowers. Opt for a substrate that is soft and loose, allowing them to exhibit natural behaviors. Substrates such as coconut coir, cypress mulch, or a mix of topsoil and play sand provide the ideal texture for burrowing.

4. Thermoregulation Support:

The substrate plays a role in thermoregulation for your Horsefield Tortoise. It should retain heat well in the basking area while allowing for effective cooling in other parts of the enclosure. A substrate with good thermal properties aids in maintaining the tortoise's optimal body temperature.

5. Safe and Non-Toxic:

Ensure that the chosen substrate is safe and non-toxic for your Horsefield Tortoise. Avoid using substrates treated with chemicals or pesticides. Natural substrates without added additives or fragrances are preferable to prevent any adverse effects on your tortoise's health.

6. Ease of Cleaning:

Maintaining a clean environment is essential for the health of your Horsefield Tortoise. Choose a substrate that is easy to spot-clean, removing feces and uneaten food promptly. Considerations such as ease of sifting or scooping contribute to the overall cleanliness of the enclosure.

7. Avoid Fine Dust:

Substrates that produce fine dust particles can pose respiratory risks for your tortoise. Fine dust may be inhaled during digging or burrowing activities. Select substrates that have minimal dust production to safeguard the respiratory health of your Horsefield Tortoise.

8. Size and Age of Your Tortoise:

The size and age of your Horsefield Tortoise influence the choice of substrate. Younger tortoises may benefit from softer substrates that are easier to burrow into, while adults can handle coarser substrates. Adjustments to the substrate can be made as your tortoise grows.

9. Substrate Depth:

Consider the appropriate depth of the substrate within the enclosure. Providing a depth of at least 4-6 inches allows for effective burrowing and digging. This depth also aids in maintaining humidity levels in a specific area if desired, such as a designated humid hide.

10. Mixing Substrates:

Combining different substrates can offer a balanced and enriched environment. For example, a mix of coconut coir and cypress mulch provides both softness and structure. Experimenting with substrate combinations allows you to tailor the enclosure to your Horsefield Tortoise's preferences.

11. Temperature Retention:

A substrate that retains heat contributes to the overall temperature gradient within the enclosure. This is especially important in the basking area, where the substrate should aid in creating a warm and comfortable spot for your tortoise to regulate its body temperature.

12. Avoid Cedar and Pine:

Steer clear of substrates made from cedar or pine, since they might emit aromatic chemicals and oils that may be hazardous to reptiles. Stick to tortoise-safe substrates like cypress mulch, coconut coir, or organic topsoil to preserve the well-being of your Horsefield Tortoise.

13. Natural Appearance:

Choosing a substrate that mimics the appearance of the natural environment boosts the overall aesthetics of the enclosure. Natural-looking substrates, such as those with a sandy or earthy tone, make a visually pleasing environment for both the tortoise and the caretaker.

14. Accessibility for Eating:

Consider the accessibility of the substrate for feeding. Substrates that are excessively loose may make it tough for your tortoise to discover and consume food. Creating defined feeding areas or using substrate with a slightly firmer texture in feeding zones helps alleviate this difficulty.

15. Substrate Changes Gradually:

If transitioning to a new substrate, implement changes gradually. Abrupt changes can upset your Horsefield Tortoise. Mix the new substrate with the existing one over time until the complete transition is made. Monitor your tortoise for any symptoms of stress during this adjustment period.

16. Bioactive Substrates:

Bioactive substrates, which incorporate living organisms like beneficial bacteria and springtails, offer extra benefits. These substrates contribute to a self-sustaining ecology within the cage, aiding in waste breakdown and creating a more dynamic habitat for your Horsefield Tortoise.

17. Moisture Levels and Humidity:

While Horsefield Tortoises prefer drier circumstances, some may benefit from localized elevated humidity for shedding. Incorporate a humid hide or moist substrate in a designated place to provide options for your tortoise. Monitoring moisture levels supports the overall well-being of your pet.

18. Compatibility with Other Enrichment:

Consider the compatibility of the chosen substrate with other enrichment items in the enclosure. Substrates that enable digging and burrowing behaviors complement elements like climbing structures, shelters, and basking locations, creating a harmonic and stimulating environment.

19. Cost and Availability:

Evaluate the cost and availability of the substrate, especially if you are setting up a big enclosure or want regular substrate changes. Choosing substrates that are commonly available and moderately priced makes the continuous upkeep of the enclosure more manageable.

20. DIY vs. Commercial Substrates:

Decide whether you prefer commercially available substrates or if you are inclined towards constructing a do-it-yourself (DIY) mix. Commercial substrates frequently come pre-packaged with consistent quality, while DIY choices allow for modification based on your tortoise's individual demands.

21. Observe Tortoise Behavior:

Pay close attention to your Horsefield Tortoise's habits and preferences. Some tortoises may display a predilection for some substrates over others. Observing their interactions with the substrate helps fine-tune the cage to fit to their specific preferences.

22. Substrate for Egg Laying:

If you have a female Horsefield Tortoise, provide an appropriate substrate for egg laying. A mix of organic topsoil and sand works well for constructing a nesting place. Make sure the depth of the substrate permits adequate burial of the egg, lowering the possibility of issues associated with the egg.

23. Upkeep and Substitution:

Schedule regular maintenance and evaluate the substrate's condition on a regular basis. Over time, damage, soiling, or modifications in the tortoise's size and activity may require replacing substrates. Keeping an eye on the substrate's state guarantees a tidy and cozy living space.

24. Consulting a Veterinarian for Reptiles:

Speak with a reptile veterinarian if you're unsure of the ideal substrate for your Horsefield Tortoise or if there are any health issues with your pet. Expert veterinarians that specialize in caring for reptiles can offer customized advice based on your tortoise's unique requirements.

25. Continue to educate yourself:

Keep up with developments in substrate choices and reptile care. Ongoing education guarantees that you can adjust and give your Horsefield Tortoise the greatest possible living environment as new goods and research are developed.

To sum up, selecting the appropriate substrate for your Horsefield Tortoise is an essential part of giving it responsible and careful care. You may build a space that meets your tortoise's physical and behavioral demands by taking into account elements like drainage, safety, natural habitat conditions, and personal preferences. Your Horsefield Tortoise's general health and happiness are enhanced by routine observation and substrate alterations, which help your pet live a happy and healthy life in captivity.

Chapter 9

Getting to Know Your Horsefield Tortoise and Forming Bonds with It

It can be gratifying and enjoyable to develop a close relationship with your Horsefield Tortoise (Agrionemys horsfieldii or Testudo horsfieldii). Although they may not be social creatures by nature like dogs or cats, tortoises do develop relationships with the people who look after them. Building a relationship with your Horsefield Tortoise entails learning about their distinctive habits, engaging in meaningful activities, and fostering an atmosphere that values wellbeing and trust. We will go over a number of socialization and bonding techniques in this extensive guide to help you develop a closer relationship with your intriguing Horsefield Tortoises.

1. Recognizing the Behavior of Tortoises:

It's important to comprehend your Horsefield Tortoise's natural behavior before starting the socialization process with them. Horsefields tortoises are often lone, self-sufficient creatures. Unlike more gregarious pets, they might not actively seek out social contacts. They are still capable of developing ties based on confidence and satisfying encounters.

2. Observing Their Personal Area with Respect:

Respecting a Horsefield Tortoises personal space is an essential part of interacting with them. It's possible that tortoises don't appreciate handling or petting as much as mammals do. Refrain from using force when your tortoise approaches you, and let them come at their own leisure. Establishing trust requires being understanding of their boundaries and exercising patience.

3. Regular Presence:

You should spend regular, quality time with your Horsefield Tortoise. Take a seat close to their enclosure, converse with them in a soothing tone, and do quiet activities. By being there, you assist them get used to your business and connect it with good things that happen.

4. Feeding by Hand:

Giving your Horsefield Tortoise a hand feed is a great approach to establish a good relationship with your presence. Serve them straight from your hand with their preferred leafy greens or little bits of fruit that are suitable for tortoises. This helps them to identify your presence with food in addition to creating a favorable interaction.

5. Establishing a Pattern:

Creating a schedule can provide your Horsefield Tortoise with comfort. A regular schedule offers a sense of predictability and is frequently well-received by tortoises, which tend to respond well to routines that are in line with their natural behaviors. This can apply to feeding schedules, daily observations, or certain activities.

6. When to Explore Outside the Enclosure:

The pleasure of bonding with your Horsefield Tortoise can be improved by letting them explore a safe, supervised area outside of their enclosure. Make sure there are no possible threats and that the area is escape-proof. They are able to roam around during this exploratory period, which could promote natural behaviors.

7. Taking Note of Natural Behaviors

Take time to observe and enjoy the habits that your Horsefield Tortoise naturally exhibits. Knowing their inclinations and instincts strengthens your bond with them, whether they are exploring, sunbathing, or burrowing. Create an enclosure that encourages these organic activities in order to create a more fulfilling and rich environment.

8. Encouragement that is constructive:
Forging a stronger link with your Horsefield Tortoise can be achieved via the use of positive reinforcement. Positive reinforcement can take the shape of soft praise, a soothing voice, or a favorite treat whenever they display desired behaviors, including coming to you or reacting to cues. This strengthens the connection that exists between your presence and good experiences.

9. Steer clear of abrupt movements:
Tortoises may react negatively to loud noises or abrupt movements. Move carefully and slowly when engaging with your Horsefield Tortoise. They could get frightened by sudden movements, which could cause tension or

withdrawal. An environment that is positive and stress-free is enhanced by a calm and kind approach.

10. Developing Experiences with Comfortable Handling:
While some tortoises may grow tolerant of handling over time and with favorable experiences, not all of them appreciate it. If handling is required, make sure the procedure is painless and gentle. Make sure they are adequately supporting their body, don't move suddenly, and pay attention to any indicators of tension or discomfort.

11. Establishing Trust Slowly
It takes time to establish confidence with your Horsefield Tortoise. Let them get used to you being around at their own rate. Don't rush conversations; instead, concentrate on building strong connections. Positive, patient, and consistent interactions are the foundation of trust.

12. Interactive Improvement:

Include interactive enrichment activities in the daily regimen for your Horsefield Tortoise. This can involve giving them puzzle feeders, concealing treats in places they have to search for, or introducing them to new items. These mental exercises help them form positive connections between their surroundings and your involvement.

13. Gently Caressing and Touching:

While some tortoises may not mind being petted or stroked, some might not. Try gently touching the head or shell of your Horsefield Tortoise if it seems comfortable being handled. Always pay attention to how they respond, and if they show any signs of stress, stop.

14. Acknowledging Unique Personalities:

Like all creatures, horsefield tortoises have unique personalities. While some people could be more reticent, others might be more gregarious and inquisitive. Acknowledge and honor these specific distinctions, adjusting your strategy to fit their particular inclinations and comfort zones.

15. Safe Contacts Outside:

Giving your Horsefield Tortoise safe outside activities can be enriching, if your climate permits. In a safe outside enclosure or during a supervised outdoor exploration session, they can be exposed to various environmental stimuli, sunlight, and fresh air.

16. Establishing a Secure Haven:

Make sure the enclosure has places for your Horsefield Tortoise to hide or retreat to where they feel safe. It adds to their general sense of security and comfort to have a place they can go to when they need a break.

17. Speaking and Singing Cues:

Although they might not react to vocal signals the same way as more talkative pets, tortoises can learn to recognize your voice. During encounters, use a soothing, quiet voice when speaking to your Horsefield Tortoise. They might come to identify your voice with good things over time.

18. Keeping an eye on stress signals

Watch out for any indications of stress in your Horsefield Tortoise. Stress signals might take the form of behavioral changes, fast breathing, or retreating inside oneself. If stress is seen, evaluate the surroundings for possible stressors and make the necessary adjustments.

19. Steer clear of Overhandling:

Some tortoises may take some touch, but it's important to avoid over-handling. In general, tortoises feel more at ease in their own environment, and frequent handling might cause stress. Focus on alternative constructive types of engagement and reserve handling for necessary instances.

20. Scent Acquaintance:

To establish connections, infuse pleasant contacts with well-known fragrances. When the tortoise is eating or exploring, you can use an item of clothing that bears your scent and position it close by. This aids in acclimating them to your presence and aroma.

21. Observing Quiet Times:

Like all reptiles, horsefield tortoises need to rest occasionally. When they burrow themselves or retreat into hiding places, honor their sleeping times and do not disturb them. Providing a calm and peaceful atmosphere for them during these times enhances their general wellbeing.

22. Veterinary Treatment and Credibility:

The general level of trust your Horsefield Tortoise has in you as their caregiver is influenced by receiving regular veterinarian care. Regular veterinary checkups guarantee that their health is tracked and that any possible problems are quickly resolved. This continued attention shows how dedicated you are to their welfare.

23. Creating a Bond Through Typical Activities:

Include regular exercises that your Horsefield Tortoise participates in. Whether it's feeding, giving them new water, or keeping up with their cage, these everyday chores foster positive interactions and strengthen your relationship as a whole.

24. Observational Moments Captured:

Just observe your Horsefield Tortoise for a while. By documenting your observations, you can gain additional insight into their habits, tastes, and distinctive features.

Using a conscious approach helps you get to know and understand your reptile partner better.

25. Long-Term Dedication and Forbearance:

Developing a close relationship with your Horsefield Tortoise takes time, patience, and understanding. Since every tortoise is unique, there may be differences in the time it takes to establish rapport and trust. Think about the process in terms of steady advancement and lifelong friendship.

26. Honoring Boundaries:

Understand and be mindful of the boundaries when interacting with a Horsefield Tortoise. They may not display the same overt signals of affection as more social pets, but they can still build friendships and express affection in their own unique ways. A happy and stress-free relationship is facilitated by acknowledging and accepting these constraints.

27. Establishing a Favorable Connection with Handling:

In the event that you handle your Horsefield Tortoise, concentrate on building a favorable association. Before and after handling, give goodies, and make sure it's a calm, gentle process. They might eventually come to equate handling with good things.

28. Presenting Tortoise-Safe Friendship:

Horsefield tortoises may occasionally coexist peacefully with other tortoises. When contemplating a companion, gradually introduce people who are safe for tortoises and keep an eye on their relationships. More stimulation and social possibilities may result from this.

29. Comparing Outdoor Pursuits:

Take into consideration engaging in outdoor activities with your Horsefield Tortoise if they like exploring the outdoors. Allow them to graze or explore at their own speed by sitting close by. Spending time together outside can improve the relationship you have with your tortoise.

30. Recording the Journey of Bonding:

Tracking your progress and cherishing memories shared with your Horsefield Tortoise can be accomplished by keeping a journal of your bonding journey. Capturing intimate experiences via pictures, movies, or even just a simple notebook helps to create enduring memories.

In conclusion, getting to know and bond with your Horsefield Tortoise is a special and gradual process that calls for tolerance, comprehension, and a dedication to their welfare. You may establish a solid and long-lasting bond with these amazing reptiles by treating them with respect, fostering pleasant interactions, and providing an enriching habitat. Keep in mind that the bond may show itself subtly and that developing trust and companionship is a pleasurable process that takes time to complete.

Chapter 10

Breeding Advice: Strategies for a Successful Reproduction of Horsefield Tortoises

A fascinating but responsible activity is breeding Horsefield Tortoises, also known as Testudo horsfieldii or Agrionemys horsfieldii. Throughout the breeding season, these resilient, little tortoises from Central Asia's dry regions display unique characteristics. A thorough grasp of the special traits of Horsefield Tortoises, as well as the provision of ideal conditions for courtship, mating, and egg incubation, are all necessary for successful reproduction. This thorough guide will cover all the information you need to effectively breed Horsefield Tortoises and protect the health of the adults and hatchlings that come from them.

1. Maturity and Age:

For effective breeding, it is imperative that the male and female Horsefield Tortoises are of an appropriate age and development. The ideal age range for female tortoises is 5–7 years, while the ideal age range for males is 4–6 years. Early breeding might produce eggs with decreased viability and cause health problems.

2. Evaluation of Health:
Before beginning the breeding procedure, thoroughly evaluate the health of the breeding pair. This involves a veterinarian examination to guarantee they are in good health and do not have any underlying medical conditions. Preventing health issues in advance helps ensure a successful and stress-free breeding process.

3. Seasons and the Hibernation Process:
Seasonal cycles affect Horsefield Tortoises, and environmental cues frequently initiate their reproductive actions. Provide various seasons,

incorporating a period of hibernation or brumation, to replicate their natural conditions. Their breeding cycle is supported and their reproductive hormones are better regulated by this seasonal change.

4. Environmental Cues:

Make environmental cues that closely resemble the breeding season's natural circumstances. This covers changes in UVB exposure, daylight hours, and temperature. By modifying these environmental elements, we can let the tortoises know when it's time for courtship and mating.

5. Distinct Enclosures:

During the mating season, divide the breeding pair of Horsefield Tortoises into their own enclosure if you have more than one. This reduces the possibility of stressors and makes it possible to pay close attention to how they behave. Additionally, keeping tortoises apart avoids potentially violent confrontations that might arise during breeding.

6. Seeing the Behaviors of Courtship:

In breeding season, Horsefield Tortoises exhibit certain courtship activities. These actions could involve head nodding, circling, and soft prodding. It is possible to determine whether the tortoises are ready to mate by watching these courtship gestures. Patience is essential since not every wooing action results in a successful mating attempt.

7. Mating Patterns:

In order for Horsefield Tortoises to successfully mate, the male must mount the female. Throughout the breeding season, this procedure could take place more than once in a short while. Even while mating activities are typically instinctive, it's important to keep an eye on the proceedings and make sure that neither tortoise is showing indications of stress or discomfort.

8. Setting Up the Nesting Site:

Give the female in the enclosure an appropriate place to build a nest and lay her eggs. Sand and topsoil should be the substrate mix in the nesting area to facilitate easy digging. Make sure the substrate is deep enough to bury the eggs—roughly 6 to 8 inches.

9. Laying of Eggs and Burial:

After a successful mating attempt, female Horsefield tortoises usually lay a clutch of eggs. The female will bury the eggs in the ready-made nesting place after they are laid. Let her finish this process on her own without interfering. To keep the eggs from becoming stressed and to make sure they stay where they are supposed to, try not to disturb the nesting place.

10. Gathering and Incubating Eggs:

Once the eggs are deposited, carefully remove them if you would rather oversee and manage the incubation process. Take care to avoid turning or tossing the eggs while gathering them. Put the eggs in an incubator that meets the necessary requirements, such as a consistent humidity and temperature. Keep a tight eye on the incubation process.

11. Temperature of Incubation:

Keep the incubation temperature constant between 82°F and 88°F (28°C and 31°C). The hatchlings' gender can be affected by the particular temperature within this range. While males may be produced by colder temperatures, females are often produced by warmer temps. Maintaining a constant temperature is essential for the growth of sound embryos.

12. Substrate for Incubation:

Use a vermiculite and water mixture or any appropriate incubation substrate for the eggs. The right amount of humidity is provided by this substrate for healthy egg development. Keep an eye on the moisture content to avoid dehydration or high humidity, which can both harm the eggs' capacity to survive.

13. Time of Incubation:

Eggs laid by Horsefield Tortoises normally take between sixty and ninety days to hatch. Throughout this time, keep a close eye on the eggs and look for any anomalies or indications of mold or fungus. If eggs seem discolored or smell strange, there can be a problem and it needs to be fixed right away.

14. Setting Up a Hatchling Enclosure:

As soon as the eggs hatch, give the hatchlings a separate enclosure. The right substrate, hiding places, basking spaces, and UVB illumination access should all be included in this enclosure. Make that the surroundings, particularly the temperature gradients, are appropriate for the hatchlings' unique requirements.

15. First Hydration and Feeding:

Fresh water for drinking and bathing should be available to hatchling horsefield tortoises. Provide a diet high in veggies, leafy greens, and calcium supplements. To encourage their growth and development, keep an eye on their feeding habits and modify the food as necessary.

16. The social dynamics of young animals

Examine the social dynamics of your hatchlings if you have more than one. Although Horsefield Tortoises prefer to live alone, at first they may put up with the company of other hatchlings. But it's important to keep an eye out for any indications of tension or hostility, and be ready to take them apart if needed.

17. Monitoring Development and Growth:

Observe the hatchlings' growth and development with great care. Note their dimensions, weight, and overall health. Frequent veterinarian examinations are necessary to guarantee the hatchlings' wellbeing and to quickly address any health issues.

18. Sustaining Appropriate Diet:

Give the breeding adults and the hatchlings a meal that is nutritiously sound and well-balanced. In particular, calcium is necessary for the production of robust shells. To create a diet that is suitable for Horsefield Tortoises at various life phases, speak with a veterinarian that specializes in reptiles.

19. How to Stop Inbreeding:

In order to avoid inbreeding, closely monitor pairings if you intend to produce Horsefield Tortoises across several seasons. Preserving genetic variation and lowering the likelihood of health problems in subsequent generations can be achieved by maintaining thorough records of ancestry and avoiding reuniting closely related individuals.

20. Improving the Environment for Hatchlings:
Give hatchlings opportunity for natural behaviors and exploration to enhance their habitat. Add objects to the substrate that have different textures, shallow water dishes, and tiny hiding places. The hatchlings' general wellbeing and behavioral development are enhanced by environmental enrichment.

21. Getting Out of the Incubation Environment:
As the hatchlings grow, gradually remove them from the regulated incubation environment. This entails bringing them up to the standard enclosure's humidity and temperature. As the environment changes, keep an eye

on how they react, and adjust as necessary to support their health.

22. Cleaning and Hygiene:

Ensure that the surroundings are tidy and sanitary for the hatchlings as well as the breeding adults. Keep feeding places, water dishes, and cages clean on a regular basis to avoid the growth of parasites and bacteria. Maintaining the general health and welfare of the tortoises depends on good hygiene.

23. Continue to educate yourself:

Keep up with developments in veterinary medicine, breeding techniques, and reptile rearing. Continual learning guarantees your ability to adjust and give Horsefield Tortoises the finest care possible throughout their whole existence, from breeding to caring for hatchlings.

24. Participation in the Community:

Interact with the reptile community and ask seasoned breeders for guidance or insights. Participating in talks with educated people, visiting reptile exhibits, and joining forums can all offer insightful viewpoints and helpful advice for effective Horsefield Tortoise breeding.

25. Appropriate Placement and Ownership:

Think about ownership responsibilities and the possibility of hatchling placement or rehoming before starting a breeding operation. Make sure you have a responsible ownership plan in place, which can involve working with respectable reptile hobbyists or locating acceptable homes for the hatchlings.

26. Legal Aspects to Take into Account:

Recognize and abide by all laws and rules pertaining to the ownership and breeding of Horsefield Tortoises. There can be regional regulations, licenses, or limits on the selling and breeding of reptiles. Ethical breeding procedures need comprehension of and adherence to these rules.

27. Maintaining Records:

Preserve thorough documentation on the breeding procedure, encompassing information on mating dates, egg laying dates, incubation conditions, and hatchling specifics. Precise documentation facilitates tracking the effectiveness of breeding endeavors, overseeing the well-being of the tortoises, and supporting conscientious breeding methodologies.

28. Variations in Genetics:

Think about how crucial genetic variety is to breeding initiatives. To avoid inbreeding depression, avoid reuniting closely related people in pairs. The general well-being and vigor of the Horsefield Tortoise population kept in captivity is influenced by the preservation of a broad genetic pool.

29. Veterinary Assistance:

Form a rapport with a veterinarian who specializes in reptiles and cares for tortoises. Frequent veterinarian examinations help to discover health problems early and guarantee that hatchlings receive the proper treatment, particularly during the breeding season.

30. Savoring the Trip:

The process of breeding Horsefield Tortoises involves careful observation, education, and management. Savor the experience of seeing these amazing reptiles in their natural habitat, assisting their life cycle, and improving their quality of life. The precise balancing act between science, observation, and a profound respect for these fascinating tortoises is what makes successful breeding so enjoyable.

In conclusion, meticulous preparation, an awareness of the animals' natural behaviors, and a dedication to responsible ownership are necessary for the successful breeding of Horsefield Tortoises. You may support the health and conservation of this intriguing species by being aware of the special traits of these tortoises, creating ideal breeding conditions, and actively taking part in their care from courting through hatchling maintenance. Recall that each successful breeding event is evidence of the caregiver's commitment, and that ethical breeding methods support the general well-being and sustainability of captive reptile populations.

FAQs

What is the Horsefield tortoise's scientific name?

A: Testudo horsfieldii is the scientific name for the Horsefield tortoise.

How much time do Horsefield tortoises usually live to?

A: Given the right care, Horsefield tortoises can live up to 75 or 100 years.

What is the average size of a Horsefield tortoise?

A mature Horsefield tortoise's shell length typically ranges from 6 to 10 inches (15 to 25 cm).

What is the Horsefield tortoise's natural habitat?

A: Horsefield tortoises are indigenous to Central Asia; they can be found in parts of Turkmenistan, Uzbekistan, Kazakhstan, and Russia.

What temperature range is best suited for an enclosure housing Horsefield tortoises?

A Horsefield tortoise enclosure should be kept between 75 and 85°F (24 and 29°C) during the day, with a small dip at night.

How often should I give my Horsefield tortoise a place to bask?

A: Offer a temperature-controlled basking area that ranges from 90-100°F (32-38°C) for 12-14 hours every day.

What kind of flooring works well in an enclosure for a Horsefield tortoise?

A mixture of dirt, coconut coir, and cypress mulch is an excellent substrate combination for Horsefield tortoises.

Is it possible to keep Horsefield tortoises outside?

A: If the weather permits, Horsefield tortoises may be housed outside with access to both shade and sunshine.

What kind of food is ideal for a Horsefield tortoise?

A healthy diet for a Horsefield tortoise should include weeds, high-fiber veggies, and infrequently, fruits.

Do Horsefield tortoises need to be given more calcium?

A: Yes, calcium supplements help maintain the health of the Horsefield tortoises' shells. Give them some calcium powder or a cuttlebone with their meal.

Can commercially available tortoise pellets be fed to Horsefield tortoises?

A: Tortoise pellets are acceptable, but they shouldn't be their main source of nutrition. Weeds and fresh greens are vital to their health.

How frequently should my Horsefield tortoise be bathed?

A: Giving your Horsefield tortoise a weekly bath should be plenty. Make sure the water is both comfortably warm and shallow.

Can tortoises from Horsefield hibernate?

A: It is true that Horsefield tortoises hibernate in the wild. But unless hibernation is done under strict supervision, it is not advised for tortoises kept in captivity.

How can I determine my Horsefield tortoise's gender?

A: Female Horsefield tortoises have shorter tails and a flat or slightly convex plastron, whilst males usually have longer tails and a concave plastron.

Do Horsefield tortoises require a friend?

A: Because horsefield tortoises prefer to live alone, it's best to maintain them that way to prevent stress and competition.

What kind of vegetables may Horsefield tortoises safely consume?

A: Edible weeds, clover, hibiscus leaves, and dandelion greens are among the safe plants. Steer clear of poisonous plants.

Are Horsefield tortoises able to swim?

A: Horsefield tortoises are not very good swimmers, despite their ability to swim. Make sure the water is shallow and keep an eye on them at all times.

How do I give my Horsefield tortoise UVB light?

A UVB bulb made especially for reptiles should be used. Make sure the tortoise spends ten to twelve hours each day in the light.

Can Horsefield tortoises be housed as pets or with other reptiles?

A: Because Horsefield tortoises have particular care needs that may differ from those of other animals, it is typically not advised to maintain them alongside other pets.

Do Horsefield tortoises excavate tunnels?

A: It is true that Horsefield tortoises dig burrows as a means of seeking refuge from extremely high temperatures.

What is shell rot in horses and how can it be avoided in my pet?

A bacterial or fungal infection of the shell causes shell rot. To stop shell rot, keep the surroundings clean, supply the right amount of humidity, and tend to any injuries right away.

How frequently should the enclosure of my Horsefield tortoise be cleaned?

A: Spot clean every day, getting rid of any food scraps or excrement. Every four to six weeks, thoroughly clean the substrate and replace it.

Q: Are glass aquariums suitable for housing Horsefield tortoises?

A: Because glass aquariums offer little airflow, they are not the best choice for housing Horsefield tortoises. Well-ventilated enclosures are advised.

What symptoms might a Horsefield tortoise have of a respiratory infection?

A: Wheezing, difficulty breathing, nasal discharge, and fatigue are among the symptoms. Consult a veterinarian if you notice any of these signs.

How can I build my Horsefield tortoise an appropriate outside enclosure?

A: Make sure the enclosure has both sunny and shaded sections, install robust fence, construct a shelter, and grow vegetation that is safe for tortoises.

I have a Horsefield tortoise. Can I use a heat mat in its enclosure?

A: Although they have their uses, heat mats are not the main source of warmth. Together with above heat lamps, make a basking area out of them.

If my Horsefield tortoise ceases to eat, what should I do?

A: Keep an eye out for symptoms, check the temperature, and make sure the lighting is adequate. See a veterinarian if the problem continues.

How can I teach my hand-fed Horsefield turtle to eat?

A: Practice patience and regularly offer food from your hand. The tortoise can eventually come to equate your hand with good things happening to it.

Is it possible to keep Horsefield tortoises indoors all the time?

A: Although they can be housed indoors, it is better for their health to allow them some outside time, particularly in the warmer months.

Is it possible to potty train Horsefield tortoises?

A: Potty training Horsefield tortoises differs from training other types of pets. It is essential to clean the enclosure on a regular basis.

How should I care for my Horsefield tortoise bedding?

A: To make a cozy and organic bedding, use topsoil, coconut coir, and cypress mulch in your substrate mixture.

How can I keep my Horsefield tortoises' beaks from growing too long?

A: Offer a range of foods, including some that must be chewed, and a cuttlebone for beak wear that occurs naturally.

Do Horsefield tortoises find handling enjoyable?

A: Horsefield tortoises usually do not want to be handled, however they do tolerate it. Reduce handling to reduce tension.

Is it possible to house Horsefield tortoises in groups?

A: If the enclosure is big enough to hold many individuals, Horsefield tortoises can be housed in groups as long as they are watched out for aggressive behaviors.

How can I provide my Horsefield tortoise with a humid hide?

A humid microenvironment for shedding can be created by using a hide box filled with moistened sphagnum moss.

Does the enclosure housing Horsefield tortoises need to have a heat gradient?

A: It is possible to create a heat gradient by placing a colder location in the low 70s°F (about 21°C) and a basking spot between 90 and 100°F (32 and 38°C).

Can iceberg lettuce be eaten by Horsefield tortoises?

A: Iceberg lettuce has little nutritional benefit despite not being harmful. For optimum nourishment, feed a mix of dark, leafy vegetables.

How can I stop my basking Horsefield tortoise from toppling over?

A: To avoid flipping, make sure the basking place is level and the tortoise has easy access to it.

Do Horsefield tortoises have daytime or nighttime habits?

A diurnal species, horsefield tortoises are active during the day.

How can I provide my Horsefield tortoise a secure outside enclosure?

A: To avoid waterlogging, use secure fencing, offer a shelter that is impenetrable to predators, and make sure the enclosure has adequate drainage.

Can Horsefield tortoises live in a home with other types of tortoises?

A: Since different kinds of tortoises may have distinct care needs and territorial habits, it's typically not advised to put them together.

How can I keep my Horsefield tortoise from becoming obese?

A: To prevent obesity, keep an eye on their food, steer clear of too much fruit, and offer a variety and well-balanced meal.

Are tomatoes safe for Horsefield tortoises to eat?

A: Although they have a high acid content, tomatoes can be fed in moderation. Serve ripe tomatoes as a treat once in a while.

How can I get my Horsefield tortoise to consume more leafy green vegetables?

A: To promote acceptability, provide a range of greens, chop or shred them finely, and combine them with favored dishes.

Is it possible for Horsefield tortoises to only consume store-bought food?

A store-bought diet shouldn't be the only thing a turtle consumes. For a balanced diet, include extra fresh greens, veggies, and sometimes fruits.

Can you feed store-bought cactus pads to Horsefield tortoises?

A: They can safely add store-bought nopales (cactus pads) in their diet.

Can you feed mushrooms to Horsefield tortoises?

A: Although certain types of mushrooms are safe, it's advisable to stay away from them as some can be poisonous. Limit your diet to veggies and safe greens.

Are carrots safe for Horsefield tortoises to eat?

A: Although they are abundant in beta-carotene, carrots can be fed in moderation. Feed as a treat once in a while.

Can you feed parsley to Horsefield tortoises?

A: Because of its high calcium-to-phosphorus ratio, parsley is okay to feed them on occasion but shouldn't be their main source of nutrition.

How do I give my Horsefield tortoise enrichment?

A: To keep their surroundings interesting, supply a range of objects for investigation, alter the enclosure's textures, and switch up the hiding places.

www.ingramcontent.com/pod-product-compliance
Lightning Source LLC
Chambersburg PA
CBHW071206290526
45796CB00008B/159